Progress with Oxford

Age 4-5

Shapes and and Measuring

Hello! I'm Leeta and this is Meeta.

Contents

Key

 Say the sound

 Draw

 Write

 Count

 Match

 Circle

 Colour

 Trace with pencil

 Find the sticker

 Play together

OXFORD

UNIVERSITY PRESS

Many shapes

 There are many different shapes. Trace over these shapes.

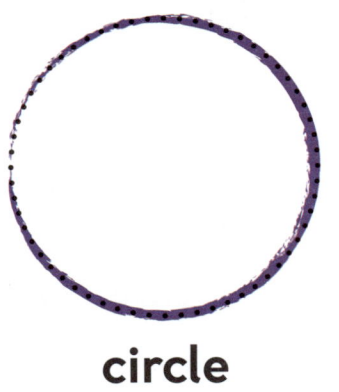

circle

square

rectangle

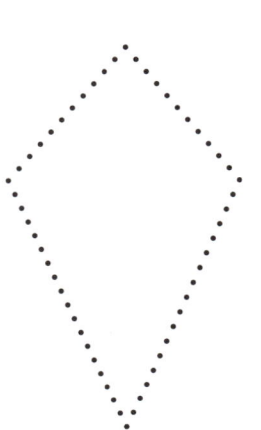

kite

triangle

What shape is a plate?

trapezium

pentagon

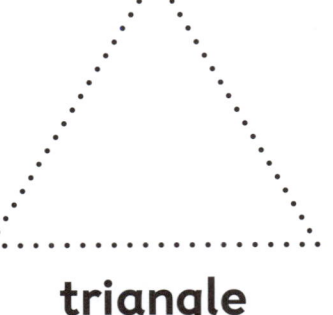

semi-circle

hexagon

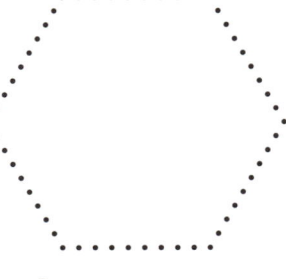

oval

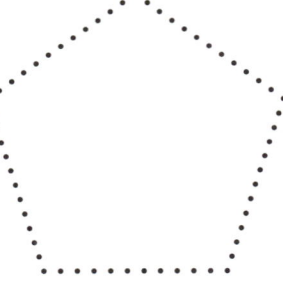

Can you say the names of the shapes on this page?

 Draw a line to match the shapes.

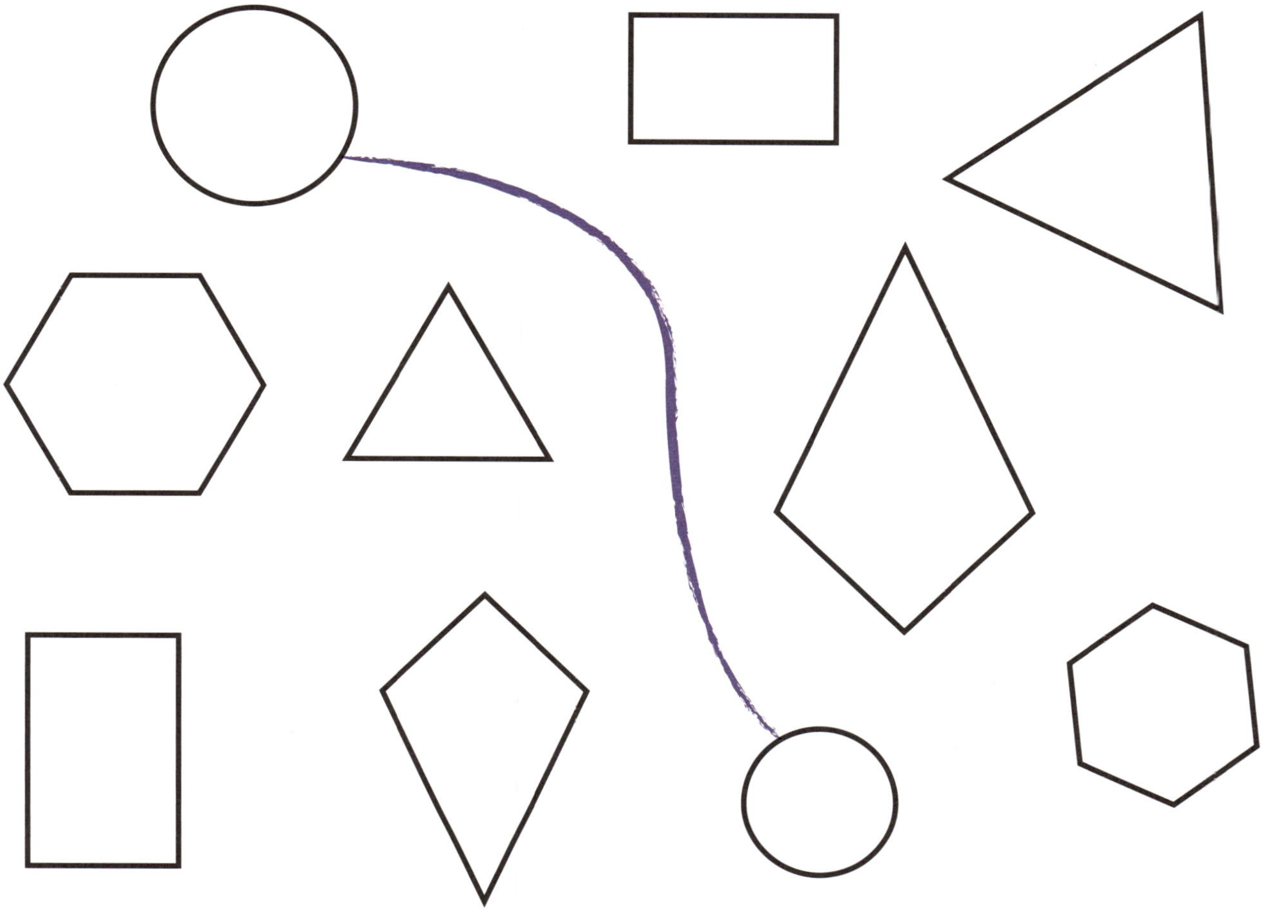

What shapes can you find around your house?

Well done!

Give yourself a sticker

Play with shapes.

Look at shapes around your house. What shape is the clock face? What shape is the television? What shape are the buttons on the remote control?

Go for a walk in the park. Find a range of leaves and compare their shapes. Use them to make a pattern.

Now – track how you're doing on page 32!

Know your shapes

 Trace these shapes.

 Draw each shape twice.

What's your favourite shape?

 Can you see the shapes in this robot? Point to each shape and say its name.

 What shape is your nose?

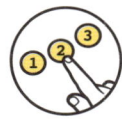 How many squares can you see on the robot?

How many triangles can you see on the robot?

 Play with shapes.

Draw shapes in the mirror after a bath.

Fill a bag with gel. Draw shapes in the gel with your finger.

Cover a board with flour. Draw shapes in the flour.

 Well done!

Give yourself a sticker

Now – track how you're doing on page 32!

Sides and corners

This rectangle has four sides and four corners.

side →

corner

Can you count the sides and the corners?

Count the number of sides.

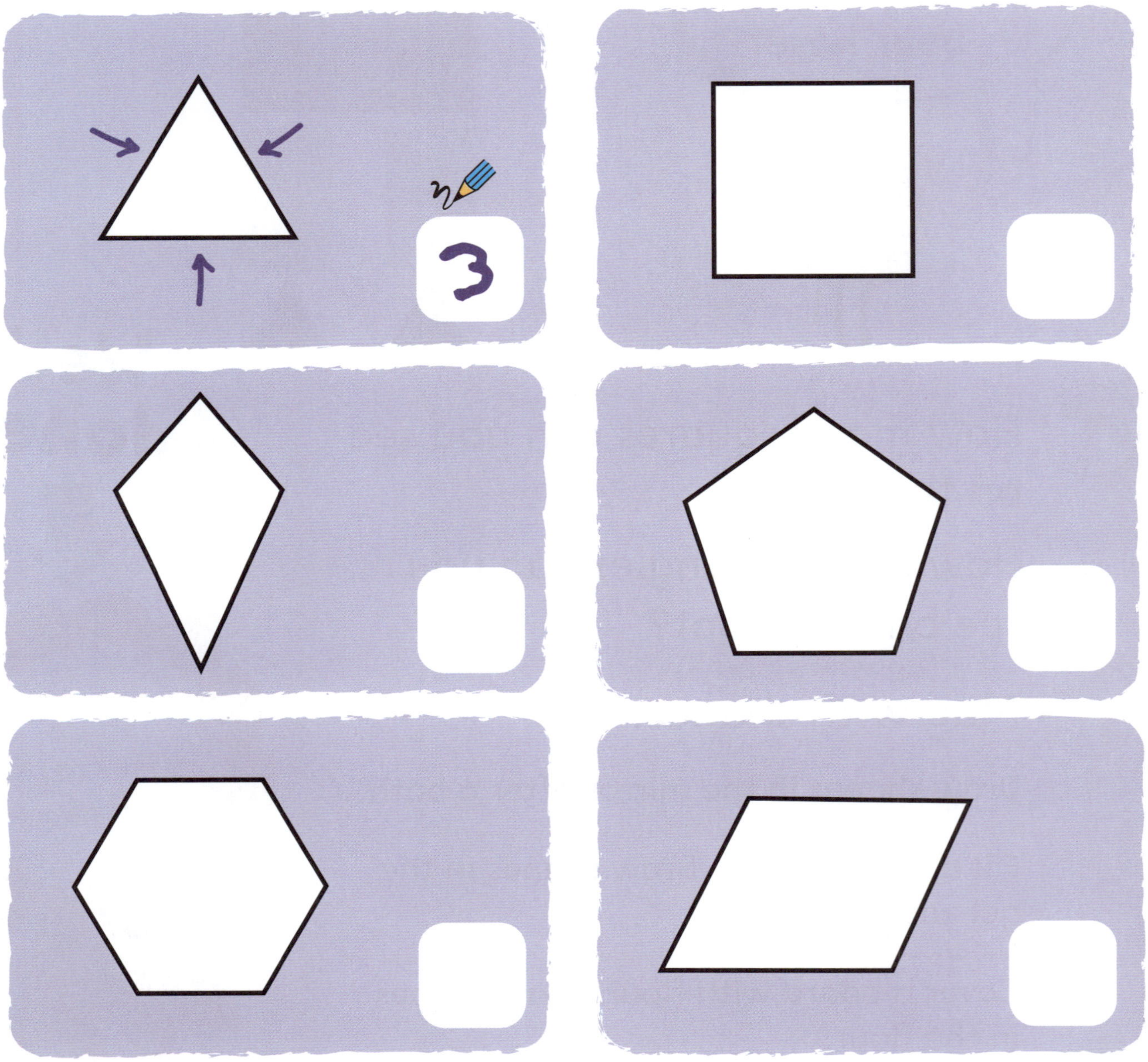

3

3

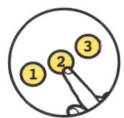 Count the number of corners.

 Put the shape stickers in the right group.

Shapes with 4 sides

Shapes with 3 sides

Shapes with 5 sides

Give yourself a sticker

Now – track how you're doing on page 32!

3D shapes

 Say the names of these 3D shapes.

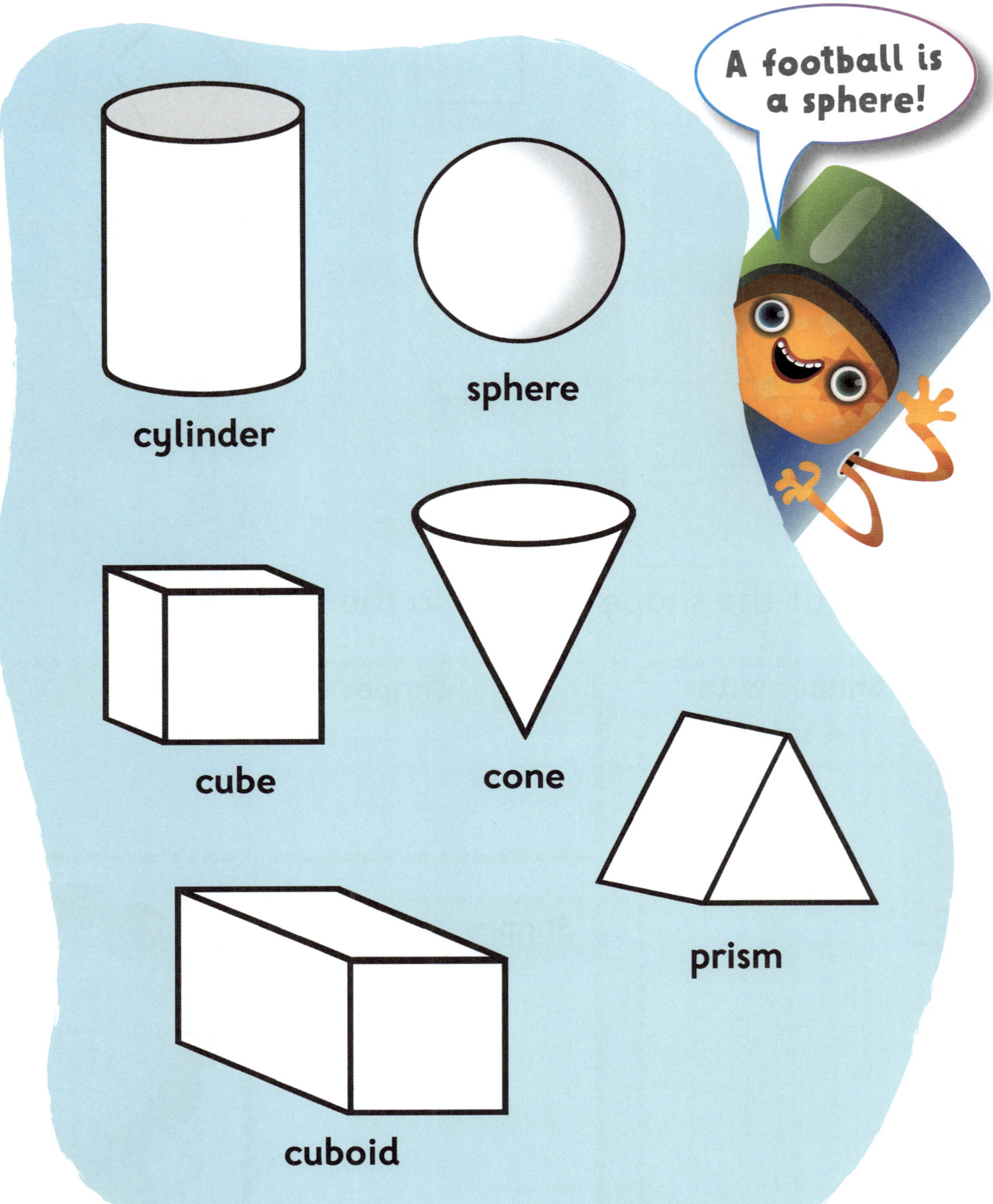

cylinder

sphere

cube

cone

prism

cuboid

A football is a sphere!

 ## Match the objects to the shapes.

BEANS

1st

Morning Flakes

Would you like to kick this sphere?

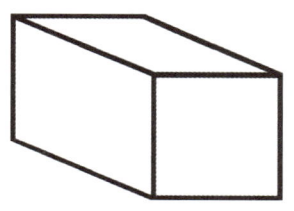

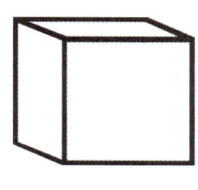

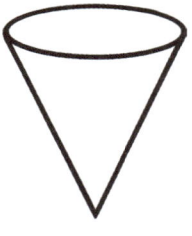

Well done!

Give yourself a sticker

 ## Play with shapes.

Use toy bricks or cardboard boxes to build a house. How will you make the roof?

Look at the bottom of a cereal box. What shape is it? Look at the bottom of a tin of food. What shape is it?

Now – track how doing on page 32!

Big and small, long and short

 Circle the biggest object in each pair.

 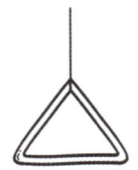

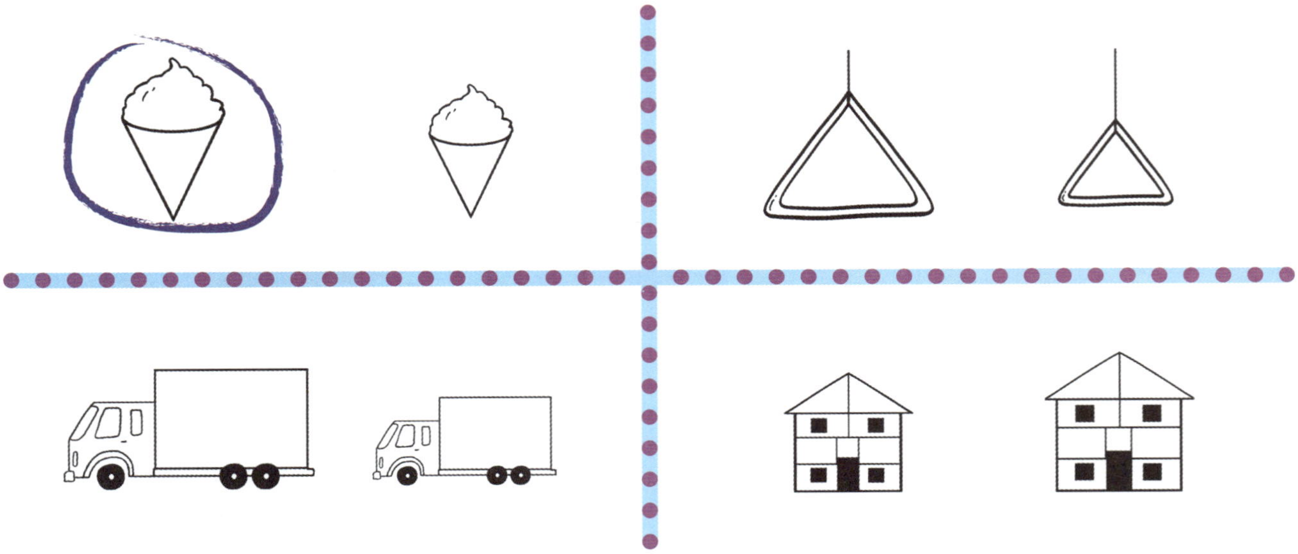

 Colour the biggest object in each row.

I am big but Meeta is small.

Long **Short**

Look at these shoes.

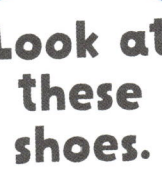

 Circle the longer object.

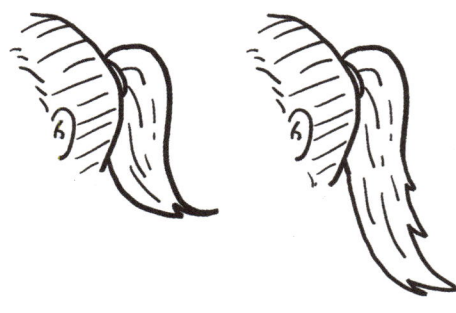

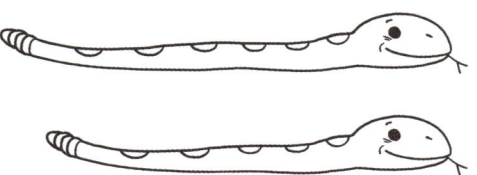

 **Play with measuring.**

Draw round the feet of everyone at home. Compare the sizes of feet. Who has the biggest feet? Who has the smallest?

Give yourself a sticker

Now – track how doing on page 32!

More or less? Full or empty?

 Point to each object in each pair and say **more** or **less**.

I want more!

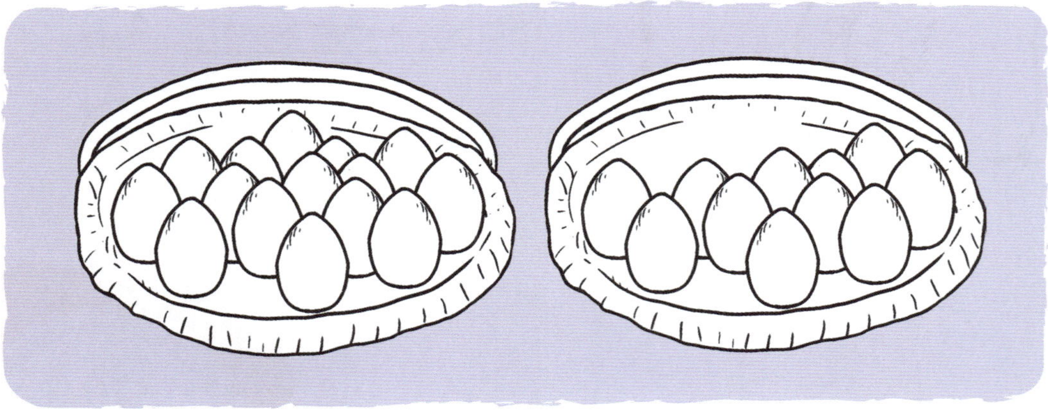

 Look how much squash is in the first jug.
Then draw **more** juice in the second jug.

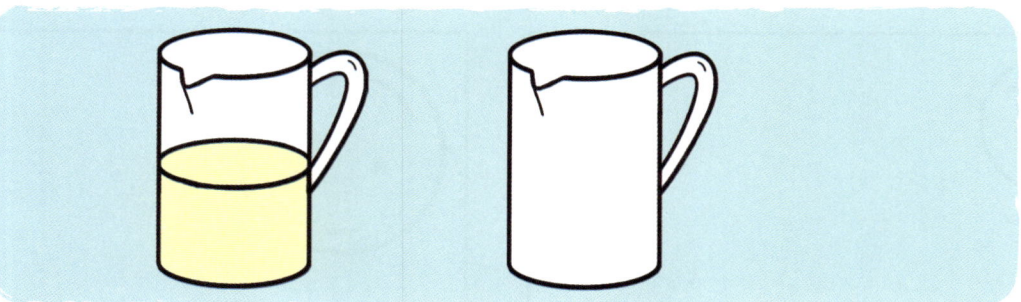

 Match the bottles to the correct words.

full

nearly full

half full

nearly empty

empty

 My glass is half empty.

Well done!

Give yourself a sticker

Now – track how you're doing on page 32!

Same size

 Sort the button stickers into the right box.

 Match the Meetas that are the same size.

Sooo many Meetas!

 **Circle the toy that is the same size.**

 **Draw a shape the same size as the one in the box.**

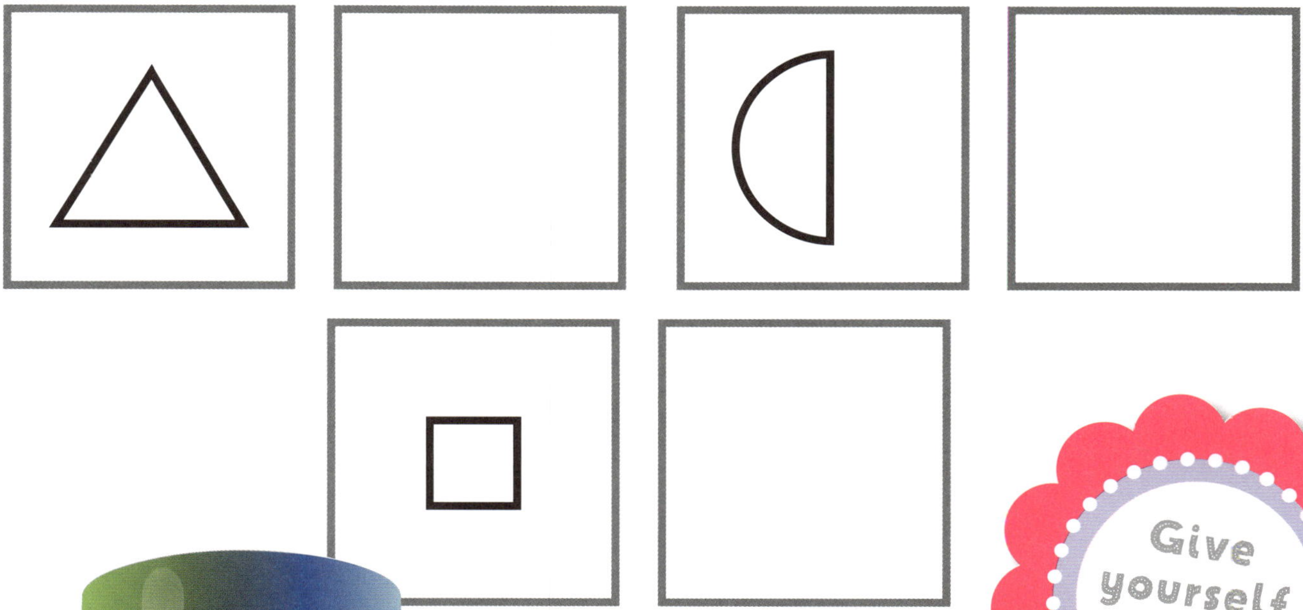

All my hands are the same size.

Give yourself a sticker

Now – track how you're doing on page 32!

How tall? How long?

 Look at the children.

How tall are you?

May	Ben	Rav	Tom	Bel

Who is the tallest child? **Bel**

Who is the tallest boy?

Who is the shortest child?

Who is the shortest girl?

 Find the stickers to order the people from the tallest to the shortest.

tallest   shortest

Stickers for page 7

Stickers for page 14

Stickers for page 16

Stickers for page 22

Character stickers

Reward Stickers

Stickers for page 23

 How long are these things?

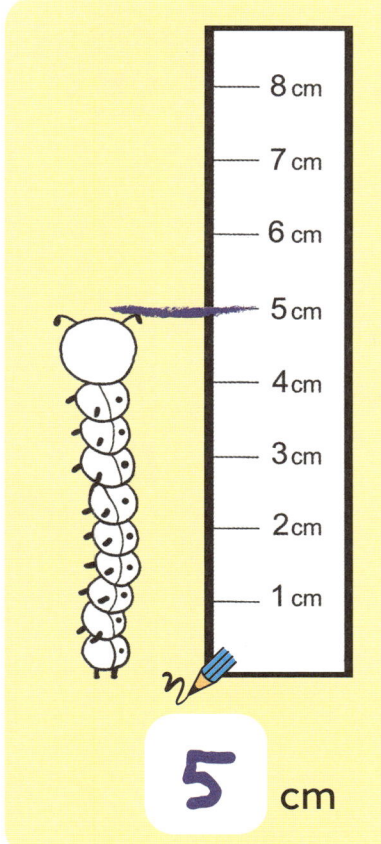

8 cm
7 cm
6 cm
5 cm
4 cm
3 cm
2 cm
1 cm

5 cm

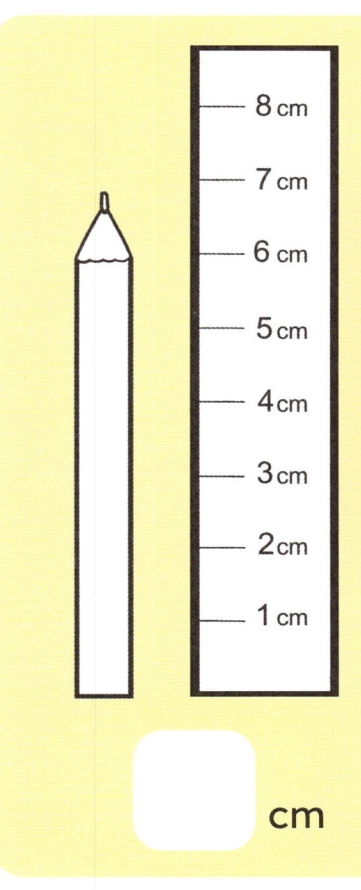

8 cm
7 cm
6 cm
5 cm
4 cm
3 cm
2 cm
1 cm

cm

Count the marks on the rulers.

I can jump half a metre!

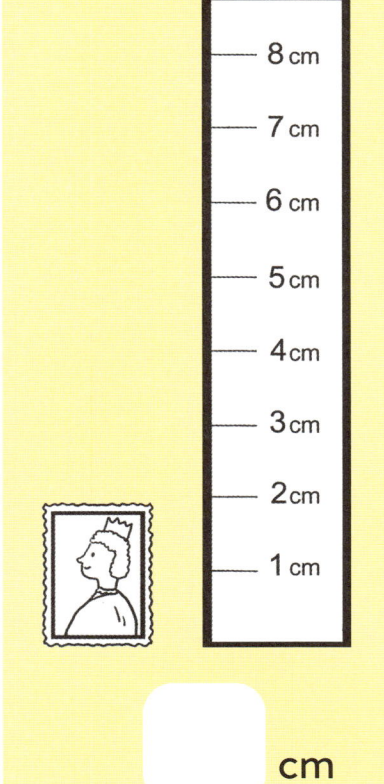

8 cm
7 cm
6 cm
5 cm
4 cm
3 cm
2 cm
1 cm

cm

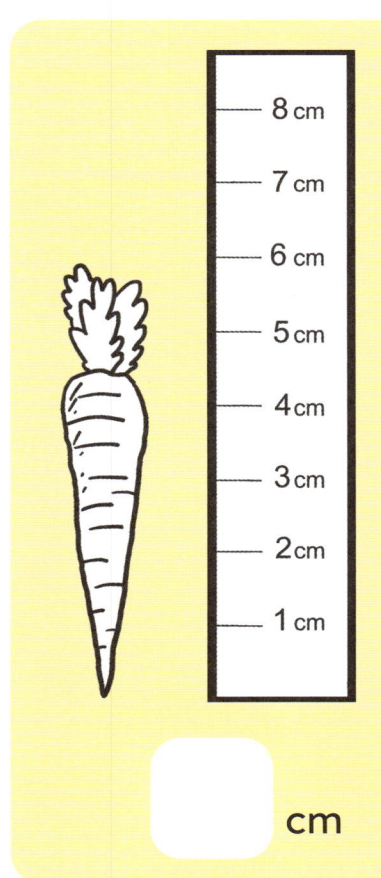

8 cm
7 cm
6 cm
5 cm
4 cm
3 cm
2 cm
1 cm

cm

Give yourself a sticker

Now – track how you're doing on page 32!

How far?

 Trace the hops.

How far have the frogs jumped?

a

4

Look at them jump!

b

c

d

 Which frog has jumped the furthest?

✎ Look at the paths.

How far do you walk to school?

a
b
c

Which path is the shortest? ☐

Which path is the longest? ☐

✎ Look at the picture.

Who has the longest walk to the park? _____

Who has the shortest walk to the park? _____

Sun

Pip

Jim

Give yourself a sticker

19   Now – track how you're doing on page 32!

Telling the time

 What happens at each of these times?

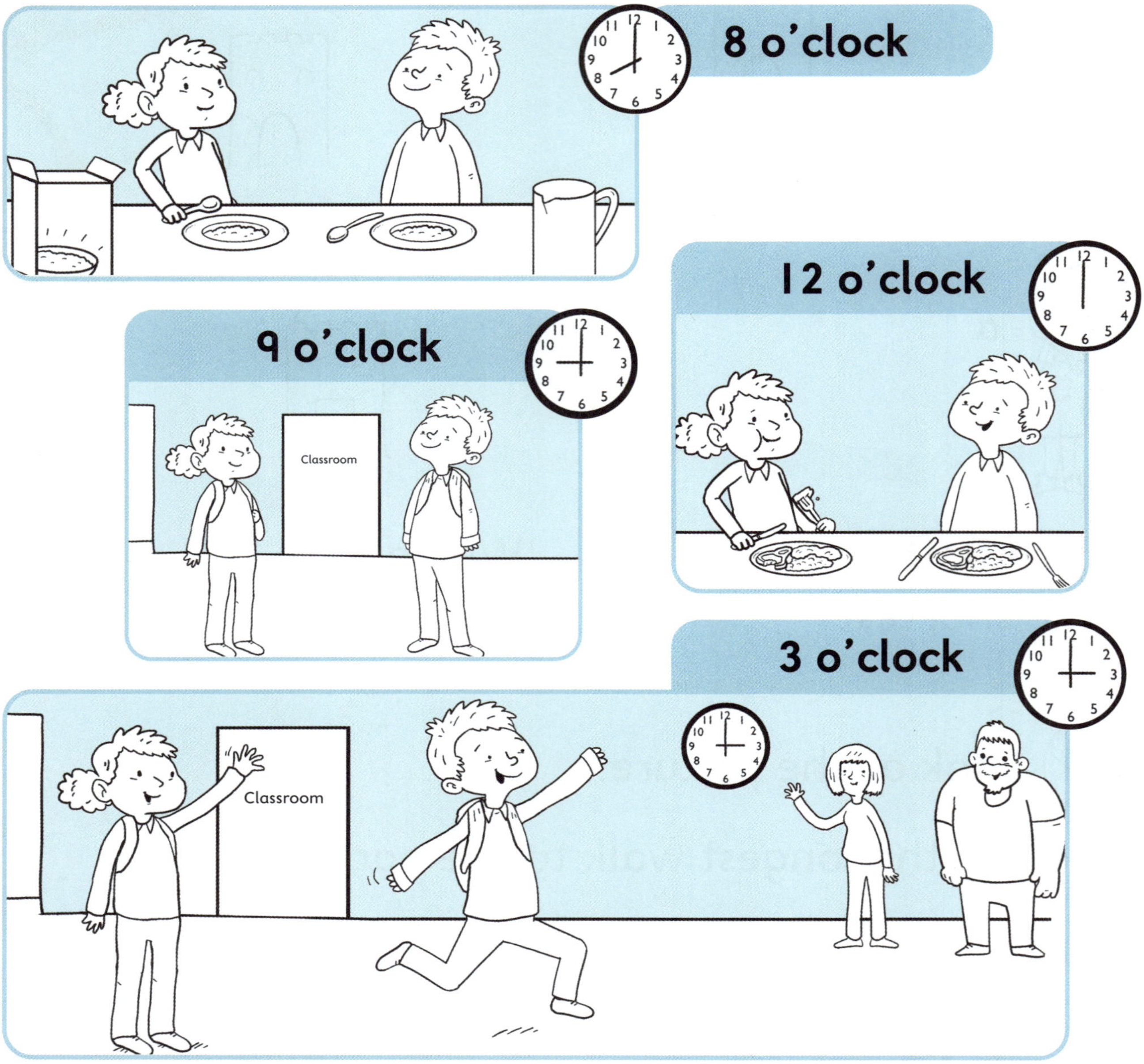

8 o'clock

9 o'clock

12 o'clock

3 o'clock

Classroom

Classroom

 What time of day do you do these things: morning, afternoon or night?

✎ Write in the numbers on the clock.

This hand shows the hour.

1

What time do you go to bed?

✎ What time is it on the clock?

☐ o'clock.

Play with measuring time.

Look at the time together. Make your own practice clock with painted stones and a stick.

Talk about the things you do every day. What time do you go to sleep? What time do you get up?

Give yourself a sticker

Now – track how you're doing on page 32!

Different coins

 Look at each coin. Say the value.

1p   2p   5p   10p

20p   50p   £1   £2

 Use the stickers to match the coins.

Counting money

Stick the right number of 1p coins next to each item.

Stick your coins here!

 3p

 1p

 2p

 3p

 2p

 **Play with money.**

Set up a pretend shop at home. Make labels and take turns to come in and buy items with the correct coins.

Sort out a pile of coins. Talk about the value of each one.

Go shopping and pay for small items with cash. Count the change together.

Give yourself a sticker

Patterns

 Say what the pattern is.

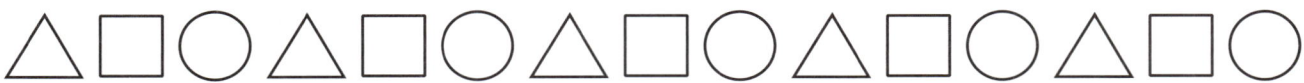

 Circle the next item in the pattern.

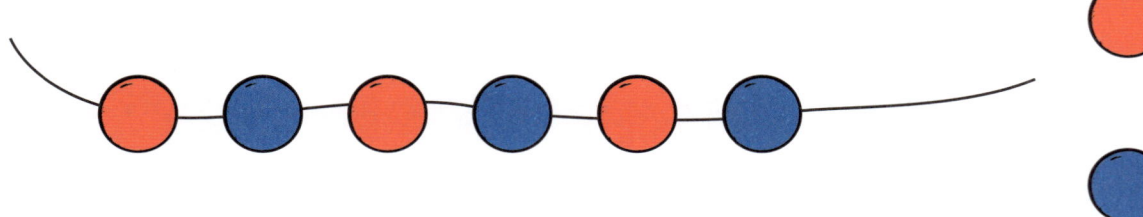

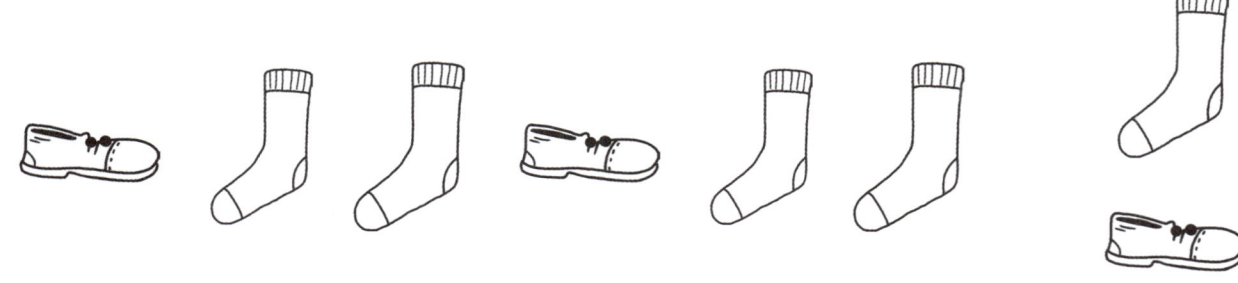

 Colour in the beads to complete the pattern.

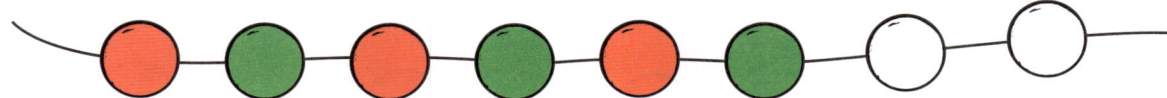

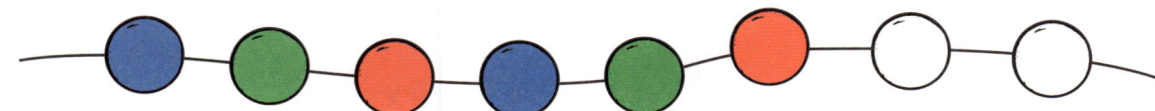

 Draw the missing pattern, shape and item.

What patterns can you see around you?

Give yourself a sticker

More patterns

 Circle the mistake in these patterns.

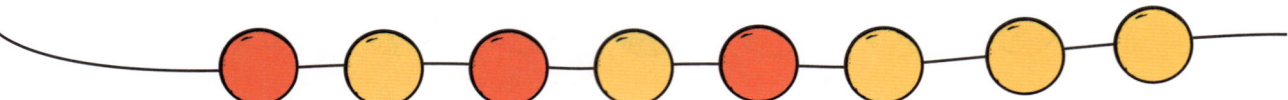

 Draw your own pattern with these items.

 Circle the patterns you can see in this picture.

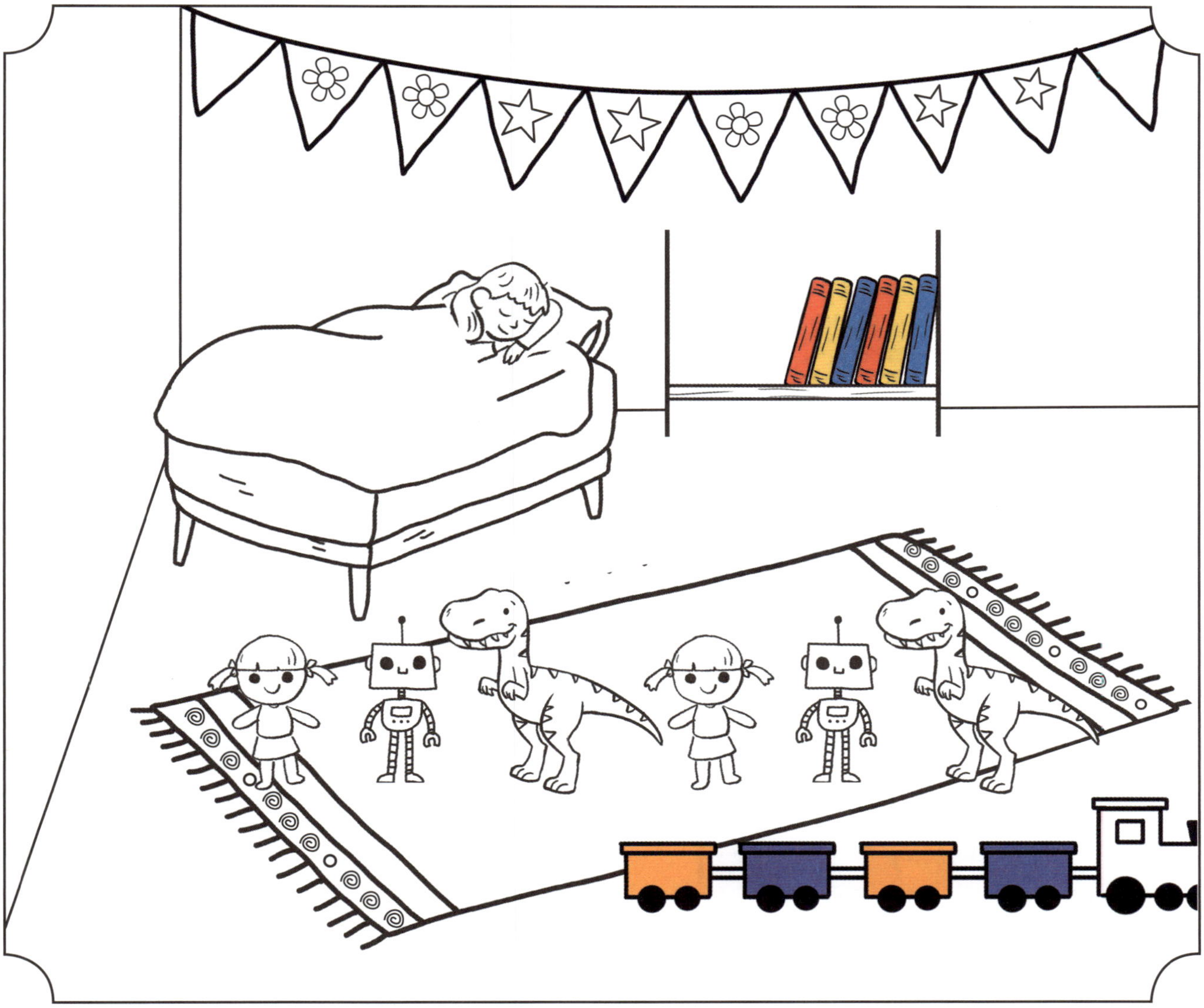

 **Play with patterns.**

Look for patterns in your home. Look at the curtains or the cushions or a rug.

Use beads of different shapes and colours to make patterns on a string.

Find different leaves in the park. Use them to make a pattern.

Give yourself a sticker

Where's Benny?

 Say where Benny is.

Is he on the box? In front of it?

 Draw Benny.

under

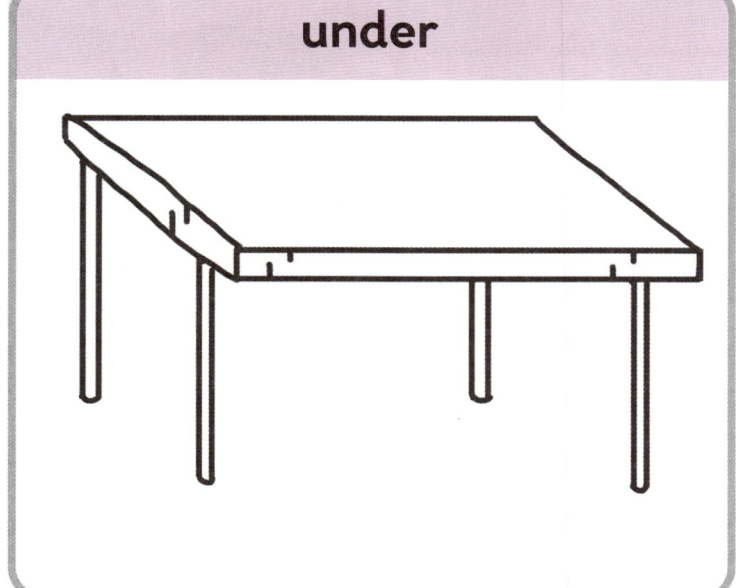

in

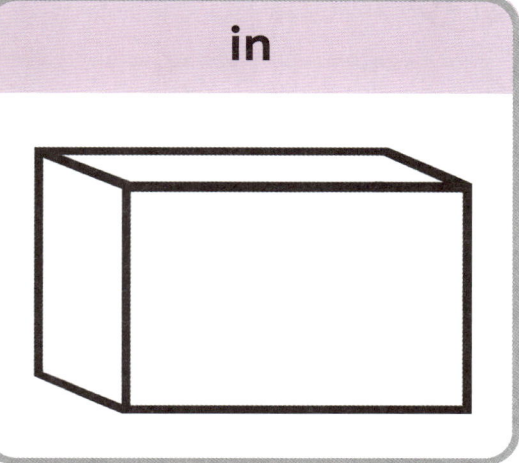

beside

on top

between

Give yourself a sticker

Now – track how you're doing on page 32!

Remember these?

 Draw a shape with three sides.

 Circle the shape with four corners.

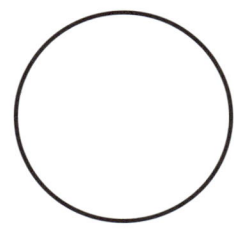 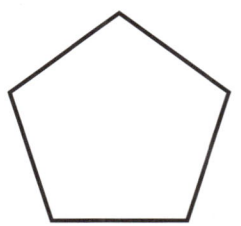 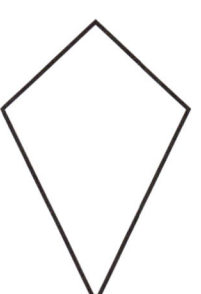 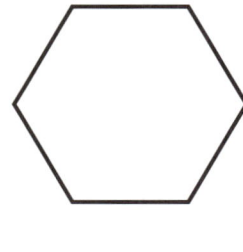

 This jug is nearly empty. Draw in the juice.

 Circle the child who has less.

Look what you know!

 **Circle the time that is different.**

 **Match the coin or coins that are the same value.**

Now – track how you're doing on page 32!

Give yourself a sticker

Progress Chart

Colour in a face.

Page	I Can . . .	How did you do?		
2–3	I can match similar shapes.	😊	😐	🙁
4–5	I can name different shapes.	😊	😐	🙁
6–7	I can count the sides and corners of a shape.	😊	😐	🙁
8–9	I can name 3D shapes.	😊	😐	🙁
10–11	I can compare big and small, long and short.	😊	😐	🙁
12–13	I can compare more and less, full and empty.	😊	😐	🙁
14–15	I can match things that are the same size.	😊	😐	🙁
16–17	I can measure how tall or how long something is.	😊	😐	🙁
18–19	I can say how far.	😊	😐	🙁
20–21	I can say when things happen in a day.	😊	😐	🙁
22–23	I can say the names of the coins.	😊	😐	🙁
24–25	I can identify patterns.	😊	😐	🙁
26–27	I can identify and draw patterns.	😊	😐	🙁
28–29	I can say where something is.	😊	😐	🙁
30–31	I can remember what I've learned.	😊	😐	🙁

How did YOU do?